Along Sandy Trails

ALONG SANDY TRAILS

Ann Nolan Clark

PHOTOGRAPHS BY ALFRED A. COHN

THE VIKING PRESS   NEW YORK

For Grandmother Kay and Jennifer

My grandmother tells me,
* "Small Papago Indian,*
* girl of the Desert People,*
* for two summer moons*
* I will walk with you*
* across the sand patches,*
* by the rock ridges*
* and the cacti,*
* through the dry washes*
* and along the sandy trails*
* that you may know the desert*
* and hold its beauty*
* in your heart forever."*

I walk with my grandmother
along a sandy trail.
The sand beneath my feet
is damp and cool
because, last night
while I was sleeping,

clouds rained down
upon our thirsty land.
Rain washed the flowers
of all the cacti,
the pincushion and the cholla,
the hedgehog and the prickly pear.

The ocotillo branches
    are tipped
    with torches
    of fire.

We sit by the trail to rest.
Nearby a giant cactus lifts
　　its many arms
　　to hold its flower crown
　　against the sky's bright blue.
Beside me a lizard's track
　　is penciled lightly
　　on the sand.
I touch it with my fingers.

A cobweb hangs between two leaves
　　like a lace mantilla
　　spun of starlight thread.

A paloverde tree
   waves its golden flowers
   above my head.
Each flower has one white petal
   and four of yellow-gold.
I ask my grandmother, "Why?"
She says, "There are some things
   no one knows."

I see a gila woodpecker
    pecking the trunk
    of a giant cactus.
If I listen . . . listen . . . listen,
    I will hear him pecking.

Under the paloverde tree
    a little ground squirrel
    sits up on his haunches
    and stores plant seeds
    in his face pockets.

Two striped squirrels
    are telling secrets
    where a fairy duster
    sways and bends.

Along the trail
    a roadrunner runs
    all stretched out
    as if he cannot get
    to where he is going
    fast enough, soon enough.
I look and see. I listen and hear.
There are so many things
    in this quiet land.

But I like best the quail.
I watch them walking,
    their black plumes bobbing
    from their red bonnets.
They walk across the trail
    near my grandmother and me,
    so busy talking together
    they do not see us.

Quail do not hop
    as some birds do.
They walk elegantly
    with quick, small steps.
Other birds walk alone,
    but quail go everywhere
    with their families
    and their friends.
They go in coveys.

My grandmother tells me,
  "Quail have many voices.
  Sometimes they murmur,
  like telling secrets.
  Sometimes they chatter
  quoit-oit-woet, quoit-oit-woet.
  They have a danger call,
  cra-er, cra-er, cra-er.
  When danger is over, they call
  qua-el, qua-el, qua-el.
  When they are lonely they cry
  kaa-wale, kaa-wale,
  and the cock quail calls
  to his mate, *yuk-hae-je."*

Above us in the sky
  a hawk soars
  in slow, wide circles.
My grandmother whispers,
  "Watch. Be still.
  See the guard quail
  sitting on the cholla,
  not eating,
  not talking,
  just sitting.
  Listen. He is calling
  cra-er, cra-er, cra-er,
  warning his covey
  of the enemy hawk."

My grandmother knows so much.
I ask her, "Is this because
    you are Indian?"
She answers me, "Perhaps,
    but also I am old,
    and with years
    comes the knowing
    of many things."

Then my grandmother says,
    "Down the trail a little distance
    I will show you something
    to remember always."
We walk along and come
    to a spreading creosote.

Under its branches, on the ground,
    in a round place
    lined with desert grass,
    is a quail's nest.

In the nest are many eggs.
One is broken.
I count them
    but do not touch them
    or make a noise of any kind.
The eggs are pinkish in color
    with brown patches,
    purple spots,
    and dots of lavender.
Mother quail must love her babies,
    she makes so beautiful
    the eggs that hold them.

My grandmother says, "Perhaps
    a snake broke one egg."
She says, "Quail put their nests
    on the ground.
    They put them in shady places
    where the bright desert sun
    cannot make them too warm.
    They know how to do this,
    but they do not know

how to make the nest safe
    from enemies."

I wish the quail would put their nests
    deep in the thorny cholla thicket
    where the cactus wrens are hatched.
But quail nest on the ground,
    not safe from anything.

When quail eggs are hatched
   the babies look
   like little round balls.
I have seen baby quail
   rolling along on the sand,
   their tiny plumes bobbing
   from their red topknots.
I wish I could see these babies
   rolling along.

I feel that I know them,
   having seen the eggs.
Newly hatched babies
   must have water at once,
   so mother quail will take them
   to drink at the water hole.
She also will teach them
   to find seeds and young plants
   and tender new leaves to eat.

Cock quail, the father,
   going first always,
   guards and protects them.
If he gives the alarm call,
   mother quail hides, watching.
Baby quail scatter
   in every direction,
   move not a feather
   until safe call is given
   by cock quail, their father.
When the sun burns the land
   quail rest in the shade
   of paloverde and mesquite
   and tall desert grasses.

At nighttime they roost
   on a low bush or tree branch,
   mother and cock quail
   guarding each end
   of the baby quail line.

I like best the quail.
But my grandmother likes
   the giant cactus,
   standing tall and stark
   against the sky.

Giant cactus gives us
　　many important things.
When it lies, a fallen giant
　　on the sun-baked sand,
　　its trunk withers and dies.
Then we use its bare, dry ribs
　　to make our house walls
　　and our fences.

The rain water stored
  in its pleated trunk
  stays our thirst
  when the winds
  of the dry moon
  sweep across our land.
Its white and yellow flower-crowns
  ripen slowly to scarlet fruit
  that we gather
  and store as food
  for the time
  of the hunger moon.

Our baskets are filled
    with the ripe fruit
    of the giant cactus
    that we have gathered,
    my grandmother and I,
    and that now we take
    to my mother's house.
The sand beneath our feet
    is deep and shifting.
The way seems long
    and our baskets heavy.
We walk and rest.
We walk and rest.

After a time of just resting,
    happy and quiet,
    Grandmother says, "Come,
    little Indian granddaughter,
    the sun travels westward
    to make the day's ending.
    Your father has worked
    his fields.
    Your mother has woven
    her baskets.
    Nighttime is waiting."

"Your father has brought firewood
   for our supper fire
   and to light the evening shadows
   that come to bring
   the dark of night."

When we have eaten supper
   and the silver moon
   shines down
   on the ashes
   of the supper fire,

I go into my mother's house
and light a store-bought candle

to remember
all our walks
these moons of summer
along the sandy trails.

IDENTIFICATION OF PLANTS

Page 8, left: Pincushion cactus in blossom

Page 8, right: Cholla cactus in blossom

Page 9, top: Ocotillo in blossom

Page 9, bottom left: Hedgehog cactus in blossom

Page 9, bottom right: Prickly pear cactus in blossom

Page 10, top: Giant, or saguaro, cactus in blossom

Page 23: Giant cacti in blossom and paloverde tree

Page 24, top: Flowering giant cactus with unripened fruit

Page 24, bottom: Ripe fruit of the giant cactus

Page 25, top: Giant cactus branch with fruit

Page 27: Ocotillo branches against the sunset